W9-AAE-158

the

ILLUSTRATED
HISTORIES

of

EVERYDAY
INVENTIONS

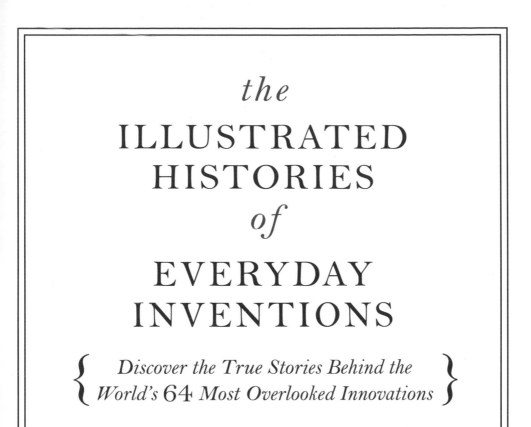

the
ILLUSTRATED
HISTORIES
of
EVERYDAY
INVENTIONS

{ *Discover the True Stories Behind the World's 64 Most Overlooked Innovations* }

by Laura Hetherington
Illustrated *by* Rebecca Pry

WHALEN
BOOK · WORKS

Kennebunkport, Maine

This book is dedicated to
Allie Hammond, the greatest
invention of all.

CONTENTS

*"There ain't no rules around here!
We are trying to accomplish something."*

—Thomas Edison, 1903

INTRODUCTION

"Invention, it must be humbly admitted, does not consist in creating out of void, but out of chaos." —Mary Shelley

Great ideas feel rare and daunting to the dreamers who seek to create something that will matter, and most importantly, last. Invention is the great Everest of this. To create something brilliant out of thin air seems nearly impossible.

But it is not thin air from which our greatest toys, technology, towers, trains, and triumphs sprang forth. It is the disorganization of daily life. An ingenious everyday invention is often quiet and always unique. It lies hidden in the monotony of daily life because of its ingenuity. It is so useful, so helpful, so fun, or so smart that it doesn't stand out in the mind, and it doesn't call attention: it does its job. It does it so well we learn to take it for granted and allow it to blend seamlessly with our lives until it becomes a part of everyday life. So let's take a moment to appreciate the trivial, brilliant, tiny, monumental things that get us all along.

In this book, you'll find the 64 everyday inventions with the most fascinating stories. I've put them in the order that they happened, so we can watch them come to life on their real timeline. Our inventions come from all over the world and dance between cultures and times as inventions cross over, are improved upon, and are shared until reaching their refined form today. The inventions we use every day come from people of all walks of life, be it the innovative minds of Mesopotamia, the pioneers and papermakers of Ancient China, or a Long Island mother who noticed pizza needed saving. Read through, have fun, and thank the toasters and the toilets in your life.

—Laura Hetherington

BEDS

THE GOOD KIND OF RESTING PLACE

IT'S HARD TO IMAGINE A TIME WHEN BEDS DID NOT EXIST, AND IT TURNS OUT THEY HAVE BEEN AROUND EVEN LONGER THAN PREVIOUSLY THOUGHT. SLEEPING MATS FROM **77,000 YEARS AGO** HAVE BEEN DISCOVERED IN THE SIBUDU CAVE ROCK SHELTER IN **SOUTH AFRICA** AND ARE THE EARLIEST EVIDENCE OF BEDDING IN HISTORY.

THE PEOPLE OF SIBUDU MADE THEIR MATS FROM COMPRESSED SEDGE STEMS, LEAVES, AND RUSHES. THEY COVERED THE BEDS IN AN OUTER LAYER OF LEAVES FROM THE CAPE QUINCE, WHICH **REPELLED INSECTS AND MOSQUITOES.**

THE INHABITANTS OF SIBUDU REGULARLY BURNED THEIR BEDS TO GET RID OF PESTS AND DECAY AND MAY HAVE ALSO USED THE MATS AS WORKSPACES. LATER ON, BEDS WERE MADE OF PLANTS, STRAW, FEATHERS, AND WATER, AND THEN THE COTTON, WOOL, AND FOAM SPRING MATTRESSES OF TODAY.

WITH THE RISING TREND TOWARDS **NATURAL, ORGANIC BEDDING** MATERIALS, IT LOOKS LIKE THE SIBUDU WERE ONTO SOMETHING.

FERMENTATION

USING BACTERIA FOR GOOD

THE EARLIEST KNOWN FERMENTED BEVERAGE IS A 9,000-YEAR-OLD MIXTURE OF FERMENTED RICE, FRUIT, AND HONEY. RESIDUE OF THE MIXTURE WAS FOUND ON POTTERY FROM THE NEOLITHIC VILLAGE JIAHU IN **HENAN PROVINCE, CHINA.**

THERE IS EARLY EVIDENCE OF BREWING AMONG THE SUMERIANS OF **ANCIENT MESOPOTAMIA,** BUT IT IS LIKELY THAT THE DISCOVERY OCCURRED LONG BEFORE, THROUGH AGRICULTURE.

THE PROCESS OF FERMENTATION WAS PROBABLY ENCOUNTERED ACCIDENTALLY WITH THE RISE OF CEREAL AGRICULTURE. **THE GROWTH OF AGRICULTURAL TECHNOLOGY MAY EVEN HAVE BEEN MOTIVATED BY THE DISCOVERY.**

- 3500 B.C. -

WHEELS

THOSE THINGS THAT ROLL

THE WHEEL WAS INVENTED IN **MESOPOTAMIA** AND MAY HAVE TAKEN A LONG TIME TO BE CREATED BECAUSE THE WHEEL DOES NOT EXIST AT ALL IN NATURE: IT IS COMPLETELY HUMAN DESIGNED.

THE FIRST WHEELS WERE **POTTERY WHEELS,** AND OTHERS WERE EVENTUALLY USED FOR CARTS AND CHARIOTS. MOST OF THESE EARLY WHEELS WERE MADE OUT OF WOOD OR STONE.

THE MOST CHALLENGING PART OF THE DESIGN WAS NOT THE WHEEL ITSELF BUT **FIGURING OUT HOW TO ATTACH IT.** THE WHEEL-AND-AXLE MODEL WAS REVOLUTIONARY AND ALLOWED THE WHEEL TO MOVE A STABLE OBJECT WHILE KEEPING IT UPRIGHT.

PUBLIC LIBRARIES

BOOKS HAVE HOUSES, TOO

SOME OF THE FIRST PUBLIC LIBRARIES WERE CREATED IN **THE MESOPOTAMIAN VALLEY.** ONE OF THE EARLIEST RECORDED LIBRARIANS WAS AMIT ANU, KNOWN AS "TABLET KEEPER" OF THE ROYAL LIBRARY AT UR AROUND 2000 B.C. LIBRARIANS HELD A PRESTIGIOUS POSITION IN SOCIETY, WERE WELL-VERSED IN LITERATURE, AND SPOKE MULTIPLE LANGUAGES.

NEW ARRIVAL

DEDICATED TO THE MUSES OF ART, KNOWLEDGE, AND SCIENCE, THE LIBRARY OF ALEXANDRIA CURATED ONE OF THE LARGEST COLLECTIONS OF SCROLLS IN THE ANCIENT WORLD AND WAS **COMPLETELY FREE AND OPEN TO THE PUBLIC.**

IN 1731, **BENJAMIN FRANKLIN** FOUNDED THE LIBRARY COMPANY OF PHILADELPHIA, ONE OF THE FIRST TO LET MEMBERS BORROW BOOKS AND TAKE THEM HOME—BUT IT REQUIRED PAID MEMBERSHIP. AFTER THE CIVIL WAR, PUBLIC LIBRARIES BECAME THE FREE RESOURCE WE TAKE FOR GRANTED TODAY. SCOTTISH-AMERICAN TYCOON **ANDREW CARNEGIE** BUILT 2,509 LIBRARIES, FIRST IN SCOTLAND AND PITTSBURGH, PENNSYLVANIA, AND THEN ACROSS THE UNITED STATES.

- CIRCA 2500 B.C. -

BATHING

YOU HATE IT TILL YOU'RE 12

AS HUMANS EVOLVED, BATHING BECAME A SOCIAL AND EVEN SPIRITUAL PRACTICE IN MANY DIFFERENT CULTURES. ONE OF THE FIRST KNOWN EXAMPLES OF BATHING ON A LARGE SCALE IS IN THE ANCIENT INDUS VALLEY CIVILIZATION OF **MOHENJO-DARO**, CIRCA 2500 B.C.

WELCOME TO INDUS VALLEY
SOUTH ASIA
We Emphasize Cleanliness!

THE GREAT BATH IN MOHENJO-DARO IS ONE OF THE EARLIEST PUBLIC BATHS IN EXISTENCE AND IS BELIEVED TO HAVE BEEN USED FOR **RITUAL BATHING.** NO EVIDENCE OF TEMPLES, PALACES, OR RULERS HAVE BEEN FOUND IN THE CITY, BUT **MOST HOMES HAD BATHS, WELLS, AND DRAINAGE SYSTEMS, SUGGESTING AN EMPHASIS ON CLEANLINESS.**

INDOOR PLUMBING AROSE IN ANCIENT GREECE AND ROME WITH THE INNOVATION OF AQUEDUCTS, SHOWERS, AND GRAND PUBLIC BATHS. **ANCIENT EGYPTIANS SPECIALIZED IN PASTES AND OILS FOR CLEANING,** AND EVIDENCE OF DAILY BATHING AMONG ANCIENT INDIANS HAS BEEN FOUND IN HINDU TEXTS. TODAY, YOU CAN TAKE A DIP IN A COMMUNAL POOL EVERYWHERE, FROM ICELAND TO JAPAN.

OILS

PASTE

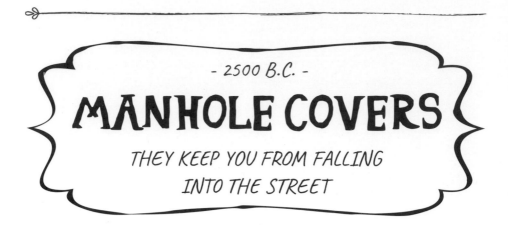

MANHOLE COVERS

THEY KEEP YOU FROM FALLING INTO THE STREET

THE CONCEPT OF MANHOLE COVERS STARTED AROUND THE TIME OF MOHENJO-DARO IN 2500 B.C., WHERE THEY USED REMOVABLE BRICKS AND SLABS OF STONE TO COVER THEIR DRAINAGE SYSTEM. MANHOLE COVERS MADE THE FIRST SEWERS SANITARY BY **PREVENTING DIRTY WATER FROM FILLING THE STREETS EVERY TIME IT RAINED.**

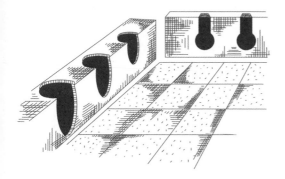

THE ROMAN EMPIRE ALSO USED **VENTED STONES** TO COVER THEIR AQUEDUCTS, AND MANHOLE COVERS EVOLVED FROM THERE.

MANHOLE COVERS ARE ROUND BECAUSE **THEY CAN'T FALL INTO THEIR CIRCULAR OPENING FROM ANY ANGLE,** WHEREAS A RECTANGULAR OR SQUARE LID COULD, IF TURNED DIAGONALLY. THE ROUND SHAPE HELPS WITHSTAND THE COMPRESSION OF THE SURROUNDING EARTH AND MAKES THEM EASY TO ROLL WHEN THEY NEED TO BE MOVED.

SCISSORS

ROCK AND PAPER COULD NEVER

SCISSORS ARE THOUGHT TO HAVE BEEN INVENTED BY **ANCIENT EGYPTIANS** AROUND 1500 B.C. TWO BLADES MADE OF BRONZE WERE CONNECTED BY A CURVED STRIP OF METAL THAT WOULD FUNCTION AS A SPRING, BRINGING THE BLADES TOGETHER WHEN SQUEEZED.

IT WAS AROUND 100 A.D. THAT THE ROMANS INVENTED THE CROSS-BLADE, PIVOTED SCISSORS THAT SERVE AS THE MODERN MODEL. THEY WERE USUALLY CONSTRUCTED OF BRONZE OR IRON AND WERE SOMETIMES **ADORNED WITH ART DESIGNED FOR THE MEETING OF THE BLADES.**

ROBERT HINCHLIFFE OF CHENEY SQUARE, LONDON, WAS THE FIRST TO MAKE SCISSORS WITH STEEL IN 1761, INITIATING THEIR FIRST MASS-PRODUCTION. SCISSORS PRODUCTION BOOMED, AND NOW WE HAVE A VARIETY OF TYPES, FROM SEWING SCISSORS TO PINKING SHEARS TO THE ROUNDED SAFETY ONES USED IN ELEMENTARY SCHOOL.

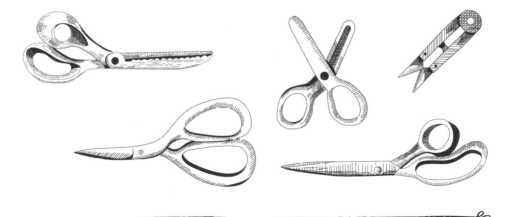

SPATULAS

THE PAST TENSE OF SPITULAS

THE SPATULA HAS BEEN FLIPPING SINCE ANCIENT ROME, AND AN EVEN EARLIER FORM LIKELY EXISTED IN **ANCIENT EGYPT** AND BABYLONIA. BUT **THERE IS NO RECORDED DISCOVERY OF THE SPATULA AS KITCHEN UTENSIL—** INSTEAD, ITS FIRST KNOWN USE WAS SURGICAL SPATULAS AND PROBES MADE OF BRONZE.

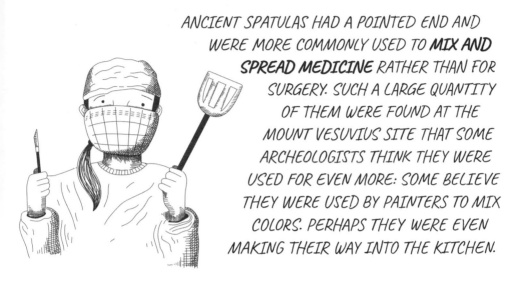

ANCIENT SPATULAS HAD A POINTED END AND WERE MORE COMMONLY USED TO **MIX AND SPREAD MEDICINE** RATHER THAN FOR SURGERY. SUCH A LARGE QUANTITY OF THEM WERE FOUND AT THE MOUNT VESUVIUS SITE THAT SOME ARCHEOLOGISTS THINK THEY WERE USED FOR EVEN MORE: SOME BELIEVE THEY WERE USED BY PAINTERS TO MIX COLORS. PERHAPS THEY WERE EVEN MAKING THEIR WAY INTO THE KITCHEN.

MEDICAL TOOLS RECOVERED FROM THE SITE OF THE ERUPTION OF MOUNT VESUVIUS IN 79 A.D. LIKELY EXISTED LONG BEFORE, AS THERE WAS LITTLE ADVANCEMENT IN SURGICAL INSTRUMENTS FROM THE TIME OF HIPPOCRATES IN 5TH CENTURY B.C. UP UNTIL THAT POINT. THE INVENTION OF THIS INVALUABLE TOOL PROBABLY OCCURRED DURING THE BRONZE AGE AND FLOURISHED DURING THE IRON AGE.

- 600 B.C. -

ASPHALT

CHAMPION KNEE SCRAPER

THE EARLIEST KNOWN USE OF ASPHALT TO BUILD ROADS WAS IN **ANCIENT BABYLON** AROUND 600 B.C. IT WAS USED TO CEMENT BRICKS, SEAL BASKETS, AND WATERPROOF BATHS AND AQUEDUCTS THROUGHOUT ANCIENT GREECE, ROME, EGYPT, AND MESOPOTAMIA.

ASPHALT WAS USED FOR PAVING THROUGHOUT EUROPE, PARTICULARLY IN ENGLAND AND FRANCE. IN 1870, A BELGIAN MAN IN THE UNITED STATES NAMED **EDWARD DE SMEDT** INVENTED A MIX MODELED AFTER NATURAL ASPHALT PAVEMENT IN FRANCE.

THE FIRST SHEET OF MODERN ASPHALT PAVEMENT WAS LAID IN 1870 IN NEWARK, NEW JERSEY OUT OF SMEDT'S FORMULA. *ITS HIGH DENSITY AND DURABILITY PROVED ASPHALT'S VALUE AS A LASTING TYPE OF ROAD.*

- 105 A.D. -

PAPER

BEATS ROCK

THE RECORDED INVENTION OF PAPER WAS IN 105 A.D., WHEN A COURT EUNUCH NAMED **CAI LUN** PRESENTED PAPER TO EMPEROR HE OF THE EASTERN HAN DYNASTY OF CHINA. HOWEVER, SAMPLES OF PAPER THROUGHOUT CHINA AND TIBET HAVE BEEN RECOVERED FROM A FEW HUNDRED YEARS EARLIER.

TA-DA!

THE PAPER WAS MADE BY SOAKING **HEMP WASTE** IN WATER, WHICH WAS THEN POUNDED AND COMBINED WITH FIBERS FROM **TREE BARK** AND **BAMBOO**.

IT WAS ORIGINALLY USED TO WRAP VALUABLES OR MEDICINE, BUT EVENTUALLY BECAME **THE PREFERRED WRITING SURFACE OVER EXPENSIVE SILK AND HEAVY BAMBOO.**

- 1100s -

SUNGLASSES

THE ORIGINAL COOL

IN PREHISTORIC TIMES, THE INUIT WORE GOGGLES MADE OF WALRUS IVORY. THEY HAD SLITS JUST THIN ENOUGH TO SEE OUT OF AND **PROTECT THE EYES FROM SNOW BLINDNESS AND SUN GLARE.**

IN 12TH CENTURY CHINA, SUNGLASSES WERE FIRST MADE FROM FLAT PIECES OF SMOKY QUARTZ. **THEY WERE PRIMARILY USED IN ANCIENT COURTS TO CONCEAL THE EYES AND FACIAL EXPRESSIONS OF JUDGES.** LOOKING COOL WAS A PLUS.

SCIENTISTS FINALLY POLARIZED LENSES IN THE 1930s, GIVING US PROTECTION FROM UV LIGHT. SOON, SHADES WERE FIRST MASS PRODUCED AND POPULARIZED ON THE BEACHES OF **ATLANTIC CITY, NEW JERSEY.**

- 1498 -

TOOTHBRUSHES

THANK GOODNESS

THE TOOTHBRUSH WAS PATENTED BY **THE HONGZHI EMPEROR OF CHINA** IN 1498, THOUGH IT IS BELIEVED TO HAVE BEEN INVENTED AROUND 700 YEARS EARLIER DURING THE TANG DYNASTY.

THEY WERE MADE OF COARSE **HOG BRISTLES** ATTACHED TO HANDLES MADE OF **ANIMAL BONE** OR **BAMBOO.**

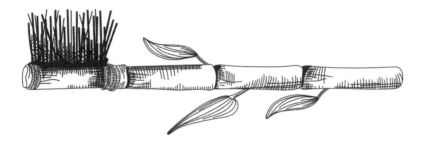

BEFORE TOOTHBRUSHES, TEETH WERE CLEANED USING **CHEW STICKS** BY THE BABYLONIANS AND EGYPTIANS. THEY WOULD CHEW ON TWIGS UNTIL THE ENDS WERE FRAYED, SOMETIMES USING STICKS FROM AROMATIC TREES TO FRESHEN THEIR BREATH.

- 1596 -

TOILETS

WHERE YOU GO TO SIT WITH YOUR PHONE FOR 20 MINUTES

THE MODERN FLUSH TOILET WAS INVENTED BY AN ENGLISH COURTIER NAMED **SIR JOHN HARINGTON** IN 1596. THE TOILET WAS A WATERPROOF BOWL ABOUT 2 FEET DEEP. IT WAS FLUSHED USING WATER FROM A RESERVOIR ABOVE, WHICH REQUIRED 7.5 GALLONS. HE CLAIMED UP TO **20 PEOPLE** COULD USE THE TOILET IN BETWEEN FLUSHES TO CONSERVE WATER.

BEFORE HARINGTON'S INVENTION, TOILETS EXISTED MOSTLY IN THE FORM OF **CHAMBER POTS, OUTHOUSES,** AND **HOLES IN THE GROUND.**

DURING THE 11TH CENTURY, SOME BATHROOMS WERE BUILT INTO CASTLES AND CALLED **GARDEROBES.** THEY WERE ESSENTIALLY HOLES POSITIONED OVER CHUTES THAT RAN STRAIGHT DOWN TO THE GROUND OR HUNG IN OPEN AIR, JUTTING OUT FROM THE SIDE OF THE CASTLE, SO THAT ALL WASTE FELL INTO THE MOAT.

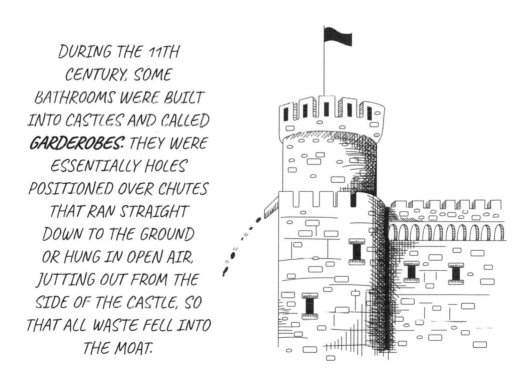

- 1743 -

ELEVATORS

INCREDIBLE FLYING CHAIRS

IN 1743, **KING LOUIS XV** HAD AN ELEVATOR KNOWN AS THE **FLYING CHAIR** BUILT IN THE PALACE OF VERSAILLES TO GIVE HIS MISTRESS SECRET ACCESS TO HIS ROOM. IT WAS ONE OF THE FIRST MODERN PASSENGER ELEVATORS AND WAS COUNTER-WEIGHTED WITH A MANUAL PULLEY SYSTEM.

EARLIER VERSIONS OF THE ELEVATOR EXISTED IN **GREECE** AND **ANCIENT ROME**, THOUGH THEY WERE MOSTLY USED AS PLATFORMS TO LIFT HEAVY LOADS OR DELIVER GLADIATORS AND ANIMALS TO BATTLE IN THE COLOSSEUM.

AFTER THE CIVIL WAR, AFRICAN-AMERICAN INVENTOR **ALEXANDER MILES** NOTICED A SHAFT DOOR LEFT OPEN DURING A RIDE WITH HIS DAUGHTER, WHICH INSPIRED HIM TO PATENT LIFE-SAVING AUTO-CLOSING DOORS.

SODA

- 1767 -

EVERYTHING'S BETTER WITH BUBBLES

*JOSEPH PRIESTLEY INVENTED **CARBONATED WATER** IN 1767 BY HOLDING A CONTAINER OF WATER OVER A FERMENTING VAT AT A BREWERY IN LEEDS, ENGLAND. AS THE BEER FERMENTED, THE CONVERSION OF SUGAR INTO ALCOHOL RELEASED CARBON DIOXIDE.*

PRIESTLEY PUBLISHED
A PAMPHLET ENTITLED
**"DIRECTIONS FOR
IMPREGNATING WATER WITH
FIXED AIR"** COMPLETE WITH A
DIAGRAM OF A CARBONATING
MECHANISM HE DESIGNED.

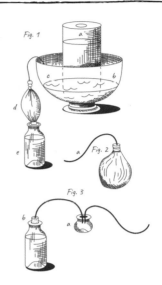

HE DID NOT MARKET THE
CARBONATED WATER AS
A SOFT DRINK HIMSELF,
BUT IT WAS HIS SCIENCE
THAT MADE OUR BUBBLY
SODA POSSIBLE. AFTER
PRIESTLEY'S DISCOVERY,
CARBONATED BEVERAGES
TOOK OFF. SOON, **CASES OF
SELTZER BOTTLES WITH
SIPHON LIDS WERE DROPPED OFF ON DOORSTEPS,** MUCH LIKE
THE CLASSIC GLASS MILK BOTTLES.

SODA-LIGHTFUL
We Deliver!

- 1767 -

WASHING MACHINES

PEOPLE DID THIS BY HAND?!

LIKE MOST THINGS, WASHING MACHINES WERE INVENTED AND IMPROVED UPON SLOWLY, OVER TIME. ONE OF THE FIRST SUCCESSFUL WASHING MACHINES WAS INVENTED BY **JACOB SCHÄFFER** IN 1767. IT WAS NON-ELECTRIC AND MANUALLY OPERATED.

HE PUBLISHED AN IMPROVED VERSION OF A WASHING MACHINE HE HAD SEEN IN *A MAGAZINE,* AND THIS MODEL WAS SO SUCCESSFUL THAT IT BARELY CHANGED FOR THE NEXT CENTURY.

SCHÄFFER ADDRESSED THE CONCERN THAT WASHWOMEN WOULD BE PUT OUT OF WORK, ASSURING THAT THEY WOULD STILL BE NEEDED TO OPERATE THE MACHINE, AS WELL AS DRY AND TREAT THE CLOTHING AFTERWARDS. *HE EVEN PUBLISHED FICTIONAL LETTERS BETWEEN TWO WOMEN DISCUSSING THE PROS AND CONS OF THE NEW CLEANING CONTRAPTION.*

BATTERIES

JUST TAKE ONE OUT OF THE REMOTE

THE BATTERY WAS INVENTED IN 1800 BY **ITALIAN PHYSICIST ALESSANDRO VOLTA.** VOLTA IS SAID TO HAVE RESISTED PRESSURE TO BECOME A PRIEST OR STUDY LAW, AND INSTEAD FOLLOWED HIS PASSION FOR PHYSICS. HE FOCUSED ON ELECTRICITY, WHICH WAS EXCITING THE SCIENTIFIC COMMUNITY AT THE TIME.

A PHYSICIST NAMED GALVANI HAD RECENTLY CLAIMED THAT WHEN TOUCHED WITH TWO DIFFERENT METALS, THE MUSCLES OF A FROG PRODUCED ELECTRICITY. HE BELIEVED HE HAD FOUND A NEW FORM OF ELECTRICITY AND CALLED IT **ANIMAL ELECTRICITY.**

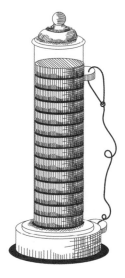

VOLTA REALIZED THAT THE TISSUE OF THE FROG WAS SERVING AS A CONDUCTOR FOR THE ELECTRICITY, RATHER THAN GENERATING THE CURRENT ITSELF. HE BUILT THE "VOLTAIC PILE," **THE FIRST WET CELL BATTERY,** BY STACKING ALTERNATING DISCS OF COPPER AND ZINC SEPARATED BY PIECES OF CLOTH OR PAPER SOAKED IN BRINE.

- 1816 -

STETHOSCOPES

THE HEART READERS

A FRENCH PHYSICIAN NAMED **RENE LAËNNEC** INVENTED THE STETHOSCOPE IN 1816, WHEN HE WAS RELUCTANT TO PLACE HIS EAR DIRECTLY AGAINST THE CHEST OF A YOUNG FEMALE PATIENT IN ORDER TO HEAR HER HEART BEAT.

HE ROLLED UP A SHEET OF PAPER INTO A MAKESHIFT TUBE, PLACING ONE END AGAINST HER CHEST, AND ONE TO HIS EAR. HE WAS ACTUALLY ABLE TO HEAR HER HEART BEAT **EVEN BETTER** THAN WITH HIS OWN EAR.

LAËNNEC'S INVENTION HELPED SHIFT MEDICAL PRACTICE FROM ITS RELIANCE ON PATIENT TESTIMONY TO THE OBSERVATIONS OF THE DOCTOR. IN 1851, THE STETHOSCOPE WAS IMPROVED TO BE BI-AURAL, AND **WHAT STARTED AS A DEVICE FOR MODESTY BECAME AN INDISPENSABLE MEDICAL TOOL.**

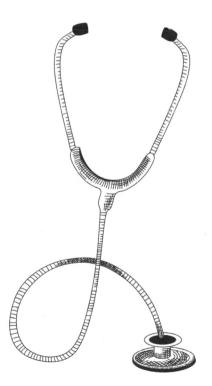

CAMERAS

CAPTURING TIME

IN 1826, AFTER YEARS OF EXPERIMENTING, **NICÉPHORE NIÉPCE** CAPTURED THE FIRST PERMANENT PHOTOGRAPH. NIÉPCE MODIFIED THE **CAMERA-OBSCURA,** WHICH USED LIGHT THROUGH A SCREEN TO PROJECT AN INVERTED AND FLIPPED IMAGE, BUT COULD NOT PRODUCE A PHYSICAL PHOTOGRAPH.

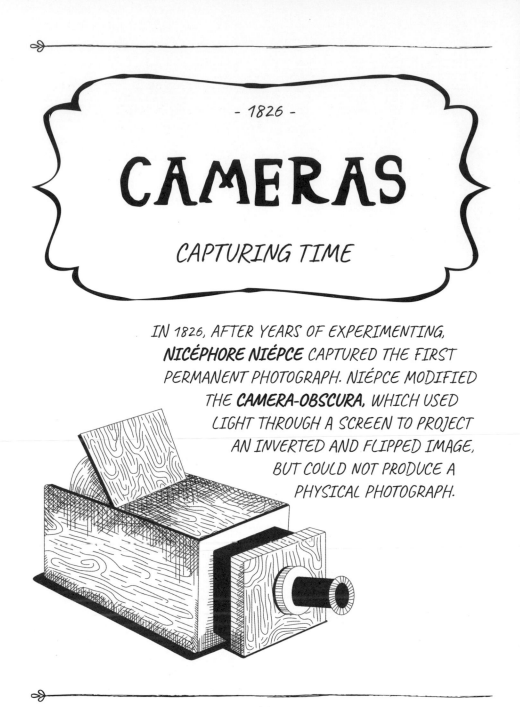

HE WAS ABLE TO PRODUCE TEMPORARY IMAGES ONTO DAYLIGHT-SENSITIVE SHEETS COATED IN SILVER CHLORIDE, BUT THESE **FADED** WITH FURTHER LIGHT EXPOSURE.

NIÉPCE EVENTUALLY COATED A PEWTER PLATE IN BITUMEN OF JUDEA, A NATURALLY OCCURRING ASPHALT. AFTER A LENGTHY EXPOSURE THROUGH THE CAMERA-OBSCURA, THE ASPHALT HARDENED WHERE IT WAS EXPOSED TO THE LIGHT. HE WIPED AWAY THE DARK AREAS WITH A SOLVENT, **LEAVING BEHIND THE FIRST PHOTOGRAPH.**

- 1834 -
FRIDGES
WHERE LEFTOVERS GO TO DIE

REFRIGERATION WAS A CONCEPT THAT ADVANCED OVER TIME, BUT IT WAS MASSACHUSETTS ENGINEER AND PHYSICIST **JACOB PERKINS** WHO CREATED THE FIRST PRACTICAL REFRIGERATOR.

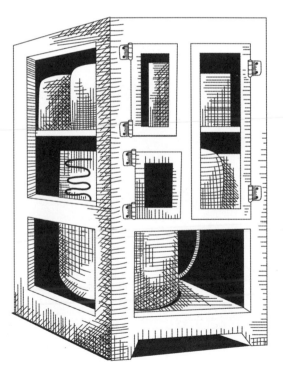

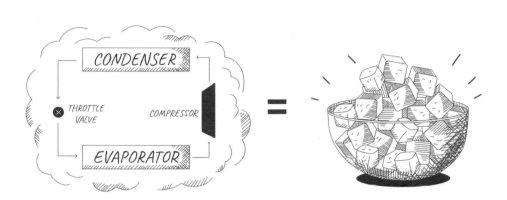

IT USED VAPOR-COMPRESSION AS ITS COOLING SYSTEM AND WAS ABLE TO PRODUCE ICE. PERKINS' INVENTION WAS MODIFIED FROM A MODEL BY **OLIVER EVANS,** WHO DESIGNED THE FIRST VAPOR-BASED REFRIGERATOR, BUT NEVER BUILT IT.

BUT THE FIRST REFRIGERATOR HAD A DANGEROUS DOWNSIDE. IT USED CHEMICALS LIKE AMMONIA TO FUNCTION, MAKING IT POTENTIALLY EXPLOSIVE. IN FACT, THERE WERE MANY FATAL ACCIDENTS INVOLVING REFRIGERATORS THROUGHOUT THE EARLY 1900s, WHICH LED TO THE DISCOVERY OF FREON AS A SAFER ALTERNATIVE. THESE DAYS, FRIDGES ARE MUCH LESS LIKELY TO **EXPLODE.**

BICYCLES

TWO-WHEELERS & BONESHAKERS

THE BICYCLE EVOLVED OVER TIME FROM A **WOODEN, TWO-WHEELED CONTRAPTION WITH NO PEDALS:** RIDERS WOULD RUN THEIR FEET ALONG THE GROUND, AND THEN PUSH OFF TO COAST.

THE FIRST VERSION OF THE MODERN BICYCLE WAS INVENTED IN THE EARLY 1860s BY A GROUP OF SEVERAL FRENCH INVENTORS AND PATENTED BY **PIERRE LALLEMENT** IN THE UNITED STATES. THEY ADDED PEDALS TO THE FRONT WHEEL, BUT THE BIKES WERE KNOWN AS **BONESHAKERS** BECAUSE OF THE RICKETY WOODEN WHEELS.

FINALLY, IN THE 1880s **JOHN KEMP STARLEY** BUILT THE UPDATED **"SAFETY BICYCLE"** FEATURING EQUALLY SIZED WHEELS, PNEUMATIC TIRES, AND A CHAIN GEAR, AND THE DESIGN HASN'T CHANGED TOO DRASTICALLY SINCE.

FOOD TRUCKS

THE SLOWPOKE'S SAVIOR

IN THE YEARS AFTER THE CIVIL WAR, A TEXAS CATTLE RANCHER NAMED **CHARLES GOODNIGHT** WANTED TO FIND A WAY TO FEED HUNGRY COWHANDS GOOD, HOT FOOD DURING LONG CATTLE DRIVES.

HE REPURPOSED AN OLD U.S. ARMY SURPLUS WAGON BY OUTFITTING IT WITH COOKING GEAR, FOOD, STORAGE COMPARTMENTS, AND SUPPLIES. AS CHUCK WAS SLANG FOR "FOOD," IT WAS DUBBED THE **CHUCK WAGON**.

IN URBAN AREAS, **PUSHCARTS** HAD BEEN USED TO HAUL FOOD, PRODUCE, AND PRETTY MUCH EVERYTHING ELSE SINCE THE EARLY 1600s. VENDORS USED PUSHCARTS TO SELL PRE-PREPARED FOOD LIKE SANDWICHES AND PIES. COMBINED, THE CHUCKWAGON AND THE PUSHCART MORPHED INTO **OUR MODERN, LATE-TO-WORK SAVIOR, THE FOOD TRUCK!**

- 1866 -

STAPLERS

PAPER FAMILY COUNSELOR

LEGEND HAS IT THAT THE FIRST STAPLER WAS INVENTED FOR **KING LOUIS XV OF FRANCE**. IT WAS SAID TO BE MADE OF **GOLD** AND **DECORATED IN PRECIOUS STONES**, THOUGH THERE IS NO REAL EVIDENCE OF ITS EXISTENCE.

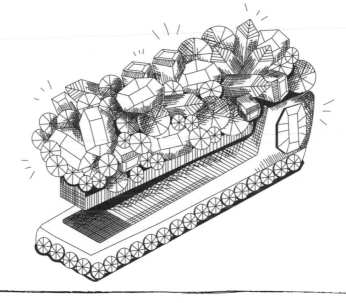

FIRST PATENTED IN THE YEAR AFTER THE CIVIL WAR ENDED, **THE FIRST MODERN STAPLER HELD ONLY ONE STAPLE AT A TIME** AND HAD TO BE RELOADED CONSTANTLY.

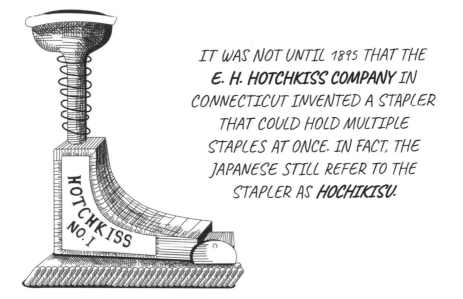

IT WAS NOT UNTIL 1895 THAT THE **E. H. HOTCHKISS COMPANY** IN CONNECTICUT INVENTED A STAPLER THAT COULD HOLD MULTIPLE STAPLES AT ONCE. IN FACT, THE JAPANESE STILL REFER TO THE STAPLER AS **HOCHIKISU**.

SNEAKERS

OUR FEET THANK YOU

THE FIRST RUBBER-SOLED SNEAKERS WERE KNOWN AS **PLIMSOLLS** AND WERE INVENTED FOR PLAYING CROQUET IN THE 1870s. THEY EARNED THE NAME **SNEAKERS** FOR BEING QUIET ENOUGH TO SNEAK AROUND IN.

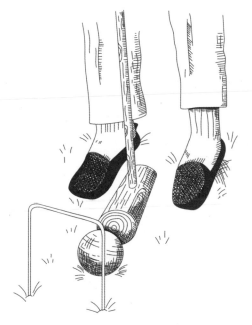

THE INVENTION OF VULCANIZED RUBBER BY **CHARLES GOODYEAR** WAS INSTRUMENTAL IN CREATING PLIMSOLLS. **VULCANIZATION** MADE IT POSSIBLE TO BOND RUBBER TO OTHER MATERIALS, LIKE FABRIC.

SAFE SHIP

THE NAME **PLIMSOLL** WAS A SAILING TERM, INDICATING A LINE ON THE OUTSIDE OF A SHIP THAT SHOWED HOW HIGH IT COULD BE LOADED WITHOUT SINKING. THIS RESEMBLED THE LINE ON THE SHOES BETWEEN THE WATERPROOF RUBBER SOLE, AND THE CANVAS UPPER. MOST SNEAKERS STILL FOLLOW THIS VISUAL DESIGN, AND WITH ATHLEISURE ON THE RISE THEY'RE TRENDIER THAN EVER.

SOGGY SHOE

CHEWING GUM

PLEASE DON'T SWALLOW IT

THE ROOTS OF CHEWING GUM LIE IN **CHICLE,** A RESIN FROM THE SAPODILLA TREE CHEWED BY THE **ANCIENT MAYA** AND **AZTECS.** ASIDE FROM FRESHENING BREATH, THE MAYANS ALSO USED CHICLE TO DEFLECT HUNGER AND THIRST.

THE AZTECS HAD **SOCIAL CODES** ABOUT CHICLE. SINGLE WOMEN COULD CHEW IT IN PUBLIC, BUT MARRIED WOMEN COULD ONLY CHEW IT IN PRIVATE.

CHICLE RECIPE FOR THE U.S.

SINGLE
- ready to -
MINGLE
- while chewing -
CHICLE

THERE WERE MANY ATTEMPTS AT SELLING COMMERCIAL CHEWING GUM IN THE UNITED STATES, BUT NONE TOOK HOLD UNTIL THE **EXILED MEXICAN PRESIDENT ANTONIO DE SANTA ANNA** SHOWED CHICLE TO **THOMAS ADAMS.** SANTA ANNA WANTED ADAMS TO PRODUCE A NEW FORM OF **RUBBER** FROM THE CHICLE. WHEN THIS WAS UNSUCCESSFUL, ADAMS BEGAN SELLING THE PRODUCT AS **CHEWING GUM.**

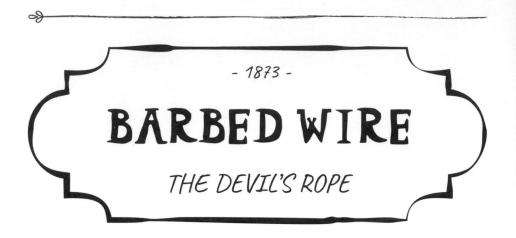

BARBED WIRE

THE DEVIL'S ROPE

AMERICAN **JOSEPH GLIDDEN** PATENTED HIS **DOUBLE-STRANDED BARBED WIRE** IN 1873. BARBED WIRE HAD BEEN DESIGNED BEFORE, BUT WAS TYPICALLY SINGLE-STRANDED. GLIDDEN'S USED TWO STRANDS OF SMOOTH WIRE IN ORDER TO HOLD THE BARBS IN PLACE, MAKING IT THE MOST EFFECTIVE, AND HIS DESIGN IS STILL AROUND TODAY.

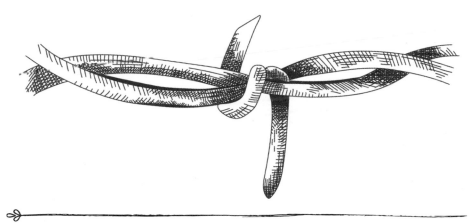

BEFORE HIS BARBED WIRE, IT WAS NEARLY IMPOSSIBLE TO **SETTLE THE WEST.** THE PRAIRIE HAD FEW TREES, MAKING ACCESS TO WOODEN FENCES DIFFICULT AND EXPENSIVE. SMOOTH-WIRE FENCES COULD BE BROKEN BY CATTLE, AND FARMERS NEEDED TO PROTECT THEIR CROPS FROM GRAZING ANIMALS:

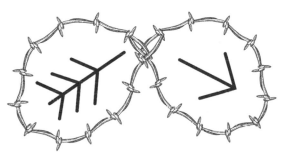

BUT BARBED WIRE ALSO AIDED SETTLERS IN STEALING LAND FROM THE NATIVE AMERICAN TRIBES IT BELONGED TO. IT WAS RIGHTLY KNOWN AS *"THE DEVIL'S ROPE."*

BLUE JEANS

LEGS' OFFICIAL FAVORITE

DENIM PANTS ARE MADE OF A THICK, TWILLED CLOTH THAT ORIGINATED IN THE PORT CITY OF **GENOA, ITALY.** THEY ARE NAMED FOR THE FRENCH WORD FOR THE CITY, **GÊNES.**

GENOA

BUT THE BLUE JEANS AS WE KNOW THEM DID NOT EXIST UNTIL AN IMPORTANT PATENT IN 1873. A NEVADA TAILOR NAMED **JACOB DAVIS** WANTED A WAY TO KEEP THE PANTS HE MADE FOR LABORERS FROM RIPPING, AND HE HAD THE IDEA TO ADD **METAL RIVETS** TO THE SEAMS. BY FASTENING METAL TO STRESS POINTS ON THE PANTS, LIKE THE SEAMS OR POCKET EDGES, DAVIS WAS ABLE TO REINFORCE THE STITCHING.

HE WENT TO **LEVI STRAUSS,** A WHOLESALE MERCHANT WHO SOLD HIM JEAN FABRIC, AND ASKED FOR THE COST OF THE PATENT IN EXCHANGE FOR A PARTNERSHIP. TOGETHER THEY MADE THE **"WAIST OVERALL,"** AND BY THE 1950s, THE MODERN BLUE JEAN WAS BORN.

PHONES

AT THIS POINT, THEY'RE AN EXTRA LIMB

*THE FIRST TELEPHONE CALL WAS FROM **ALEXANDER GRAHAM BELL** TO HIS ASSISTANT, THOMAS WATSON: **"MR. WATSON, COME HERE. I WANT TO SEE YOU."** BELL PATENTED THE TELEPHONE IN 1876, CHANGING COMMUNICATION FOREVER.*

A TEAM OF INVESTORS HAD ENLISTED BELL TO DEVELOP THE HARMONIC TELEGRAPH, A METHOD OF TRANSMITTING MULTIPLE TELEGRAMS AT ONCE TO LOWER COSTS. BUT BELL WAS MORE INVESTED IN DEVELOPING A VOICE TRANSMITTING DEVICE. HE HAD TRIED TO TEACH HIS DOG, A RESCUE TERRIER NAMED TROUVE, TO SPEAK, AND SUPPOSEDLY GOT IT TO ARTICULATE **"HOW ARE YOU, GRANDMA?"**

Give me a treat first, then I'll say your grandma thing...

AFTER BELL'S INVENTION, IT WAS ONLY A MATTER OF TIME BEFORE WE ALL WERE GLUED TO OUR SMARTPHONES. THE FIRST **TELEPHONE LINE** WAS CONSTRUCTED IN 1877, THE **ROTARY PHONE** CAME INTO USE AROUND 1890, AND **CELL PHONES** FINALLY MADE THEIR MOVE FROM CARS TO MOBILE IN 1973.

CALLING...

LIGHT BULBS

NOT JUST AN IDEA

THOMAS EDISON WAS THE FIRST TO INVENT A LIGHT BULB PRACTICAL FOR HOME USE IN 1879. PREVIOUS SUCCESSES AT GENERATING ELECTRIC LIGHT WERE EITHER SHORT-BURNING OR TOO EXPENSIVE. EDISON REMOVED OXYGEN FROM THE BULB USING A VACUUM PUMP, OPENING UP THE POSSIBILITY OF USING CARBON OVER EXPENSIVE FILAMENTS LIKE PLATINUM.

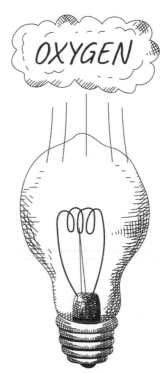

OXYGEN

EDISON'S BREAKTHROUGH CAME WHEN HE EXPERIMENTED WITH **CARBONIZED BAMBOO** AS THE FILAMENT. IN THE FIRST TEST IT BURNED FOR ABOUT 14 HOURS, AND EVENTUALLY SURPASSED 1,200 HOURS, MAKING IT THE FIRST BULB TO PROVIDE A LONG-TERM LIGHTING SOLUTION.

BUT **PERHAPS EDISON'S BIGGEST INVENTION WAS MARKETING HIMSELF AS A FAMOUS INVENTOR.** HE BRANDED ALL INVENTIONS WITH HIS NAME, PORTRAIT, AND SIGNATURE. IN THE "WAR OF CURRENTS," HIS AD CAMPAIGN CALLED NIKOLAS TESLA'S MORE POWERFUL ALTERNATING-CURRENT POWER DANGEROUS, SO THAT CONSUMERS WOULD USE THE EDISON DIRECT-CURRENT POWER SYSTEM INSTEAD.

CASH REGISTERS

THE INCORRUPTIBLE CASHIER

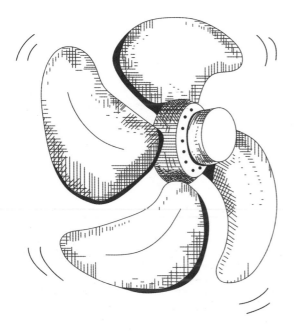

WHILE ON A VOYAGE ACROSS THE ATLANTIC, TAVERN OWNER **JACOB RITTY** NOTICED EQUIPMENT THAT COUNTED THE NUMBER OF REVOLUTIONS MADE BY THE SHIP'S PROPELLER.

HE FIGURED THAT THE SAME MECHANICS COULD BE APPLIED TO **KEEP TRACK OF SALES** AT HIS BUSINESS.

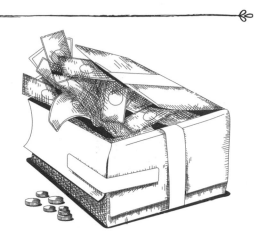

BACK HOME IN OHIO, HE DEVELOPED THE FIRST CASH REGISTER WITH HIS BROTHER, A MECHANIC NAMED **JOHN RITTY**. THEY RECEIVED THE PATENT IN 1879, AND THE FIRST COMMERCIAL MODEL WAS DUBBED RITTY'S *"INCORRUPTIBLE CASHIER."* THIS TURNED OUT TO BE MORE OF AN ASPIRATIONAL NAME, AS THE REGISTER STILL RELIED ON THE HONESTY OF THE OPERATOR TO ENTER THE SALES INTO THE REGISTER PROPERLY.

NAIL CLIPPERS

THEY'VE ONLY GOT ONE JOB, BUT IT'S AN IMPORTANT ONE

EUGENE HEIM AND **OELESTIN MATZ** OF CINCINNATI, OHIO, CREATED **THE FIRST MODERN, CLAMP-STYLE NAIL CLIPPER** IN 1881. COMPLETE WITH A DIAGRAM AND DETAILED MECHANICAL DESCRIPTION, THEIR CLIPPERS USED SPRING-STEEL AND A YOKE TO BRING THE CUTTING ENDS TOGETHER.

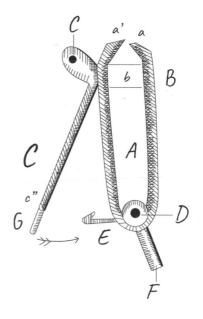

BEFORE THAT, NAILS WERE KEPT SHORT BY PARING WITH A **SMALL KNIFE** OR **SCISSORS**, SIMILAR TO HOW YOU MIGHT PEEL AN APPLE.

THERE WERE SOME **ODD SUPERSTITIONS** SURROUNDING NAIL CLIPPING IN THE LATE 1880s. AN ARTICLE BY THE NEW YORK SUN CLAIMED IT WAS BAD LUCK TO TRIM YOUR NAILS ON THE WEEKENDS. IT DECLARED THAT CUTTING THEM ON FRIDAY WAS "PLAYING INTO THE DEVIL'S HAND," "INVITING DISAPPOINTMENT" ON SATURDAY, AND CAUSING BAD LUCK FOR THE WEEK IF DONE ON SUNDAY. WHO KNEW NAILS HAD SO MUCH POWER?

- 1899 -

PAPERCLIPS

STRONG BUT TEMPORARY

OFFICE MEMO

If you are driving to work, please park your car around back.

Please do not allow any women protesters into the building. Votes for women?! Hah!

BEFORE THE PAPERCLIP, PAPER WAS HELD TOGETHER BY **THREADING A RIBBON THROUGH TWO PARALLEL CUTS** IN THE CORNER OF THE PAGES. IT WAS SOMETIMES SEALED WITH WAX. THE RIBBON ITSELF WAS SOMETIMES COATED IN WAX TO MAKE IT EASIER TO MOVE, OR TO SEAL THE ENDS FOR A STRONGER FASTENING.

LATER, **IRON STRAIGHT PINS** USED FOR TAILORING BECAME THE POPULAR METHOD. THEY MADE A SMALLER HOLE THAN RIBBONS, BUT OFTEN PIERCED FINGERS AND TENDED TO RUST.

WARNING!
May cause bleeding on documents!

MANY DIFFERENTLY SHAPED PAPER CLIPS WERE INVENTED THROUGHOUT THE 19TH CENTURY, BUT NONE WORKED WELL ENOUGH TO STICK UNTIL THE GEM PAPERCLIP BY **WILLIAM MIDDLEBROOK** IN 1899. WITH THE INVENTION OF STEEL WIRE, THEY COULD EASILY BE BENT INTO THEIR UNIQUE SHAPE AND AFFORDABLY PRODUCED, AS WELL AS CREEPILY POP UP ON MICROSOFT WORD TO GIVE YOU WRITING ADVICE.

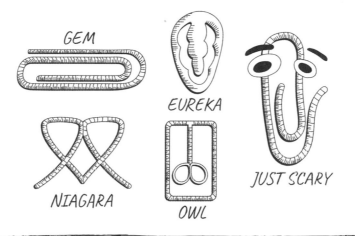

GEM

EUREKA

NIAGARA

OWL

JUST SCARY

GOT BEAR?

OUTDOORSY PRESIDENT **THEODORE "TEDDY" ROOSEVELT** WAS THE INSPIRATION BEHIND THE CLASSIC TEDDY BEAR. LEGEND HAS IT THAT ROOSEVELT WAS ON A HUNTING TRIP IN MISSISSIPPI, AND HE HAD YET TO FIND A SINGLE BEAR.

A MAN IN THE HUNTING PARTY NAMED HOLT COLLIER FOUND A **BLACK BEAR** AND TIED IT TO A TREE, SUGGESTING ROOSEVELT SHOOT IT. AN AVID HUNTER, ROOSEVELT CONSIDERED IT DISHONORABLE, AND REFUSED.

ONCE WORD GOT OUT, POLITICAL CARTOONIST CLIFFORD BERRYMAN SATIRIZED THE STORY IN A CARTOON FOR *THE WASHINGTON POST.* CANDY SHOP OWNERS AND STUFFED ANIMAL MAKERS MORRIS AND ROSE MICHTOM WERE INSPIRED TO NAME A STUFFED TOY BEAR "TEDDY'S BEAR," WHICH SOON BECAME **ONE OF THE MOST POPULAR AND ENDURING CHILDREN'S TOYS OF ALL TIME.**

WINDSHIELD WIPERS

SEEING WHILE DRIVING IS ALWAYS GREAT

IN 1902, A WOMAN FROM ALABAMA NAMED **MARY ANDERSON** WAS VISITING NEW YORK CITY. STUCK IN TRAFFIC DURING A SNOWSTORM, SHE WAS STRUCK BY HOW MUCH THE BAD WEATHER HINDERED DRIVERS.

MOTORISTS WERE EITHER **GETTING OUT TO CLEAR THEIR WINDSHIELDS** OR OPENING THE PANELS ON THEIR FRONT WINDOWS TO SEE, EXPOSING THEMSELVES AND THEIR PASSENGERS TO THE STORM.

Fig. 1

ANDERSON DREW UP A DESIGN FOR **WIPERS CONTROLLED BY A HANDLE ON THE INSIDE OF THE VEHICLE,** NOTING THAT THEY COULD BE REMOVED DURING FAIR WEATHER SO AS NOT TO OBSTRUCT THE VIEW. SHE RECEIVED A PATENT FOR HER INVENTION THE FOLLOWING YEAR.

- 1903 -

ICE CREAM CONES

WHY HAVE A BOWL WHEN YOU CAN MAKE A MESS?

THERE ARE SURPRISINGLY FIERY DEBATES REGARDING THE TRUE INVENTOR OF THE ICE CREAM CONE. IN THE LATE 1800s, ICE CREAM WAS SERVED IN GLASS DISHES THAT WERE RETURNED TO VENDORS AND REUSED. CUSTOMERS OFTEN BROKE OR WALKED OFF WITH THE CONTAINERS, OPENING THE DOOR FOR A BETTER SOLUTION.

ITALO MARCHIONY FILED AN AMERICAN PATENT IN 1903 FOR AN EDIBLE ICE CREAM CUP MACHINE AND CLAIMED TO HAVE BEEN MAKING THEM SINCE 1896. BUT, *ANTONIO VALVONA* FILED A PATENT 2 YEARS EARLIER FOR A BISCUIT-CUP MAKER FOR ICE CREAM IN ENGLAND, WHILE *ABE DOUMAR* PROMOTED HIS "WORLD'S FIRST CONE MACHINE" ON CONEY ISLAND.

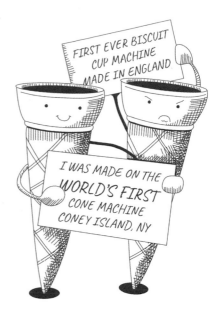

STILL DOZENS OF OTHER CLAIMS EXIST OF THE FIRST ICE CREAM CONE. MOST ORIGINATE AT THE 1904 WORLD'S FAIR; OTHERS PRESENT EARLY EVIDENCE THROUGHOUT EUROPE. BUT *THE TRUTH WILL REMAIN DESSERT'S GREATEST MYSTERY.*

REPLACEABLE RAZOR BLADES

THE ORIGINAL BAIT-AND-HOOK BUSINESS MODEL

TRAVELING SALESMAN **KING CAMP GILLETTE** INVENTED THE FIRST ACCOMPANYING SAFETY RAZOR AND DISPOSABLE BLADES. GILLETTE RECOGNIZED THE HASSLE AND EXPENSE OF TAKING A RAZOR BLADE TO GET SHARPENED EVERY TIME IT GOT DULL. HE CONCEIVED OF THINNER BLADES THAT WERE INTERCHANGEABLE AND LESS EXPENSIVE TO PRODUCE.

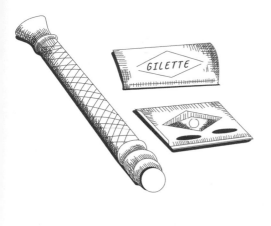

BUT THE REAL SIGNIFICANCE OF THIS INVENTION WAS THE INTRODUCTION OF THE TWO-PART PRICE MODEL. BY MAKING CUSTOMERS PURCHASE THE BLADES SEPARATELY FROM THE RAZOR, CONSUMERS ARE GUARANTEED TO CONTINUE SPENDING AFTER AN INITIAL PURCHASE, AND A COMPANY WILL CONTINUE TO PROFIT.

TODAY THE TWO-PART PRICE MODEL IS USED FOR CONSOLES AND VIDEO GAMES, PRINTERS AND INK CARTRIDGES, COFFEE MACHINES AND PODS, AND COUNTLESS MORE INDUSTRIES. **IT KEEPS CONSUMERS HOOKED BY THE EXPENSE OF SWITCHING COSTS:** TO STOP BUYING THE ACCESSORIES, YOU HAVE TO PURCHASE A WHOLE NEW EXPENSIVE BASE. THIS INVENTION REVOLUTIONIZED SALES PRACTICES AND MADE SHAVING A LOT EASIER.

WE GOT YA REAL GOOD.

T-SHIRTS

BUTTONS? WHO NEEDS 'EM

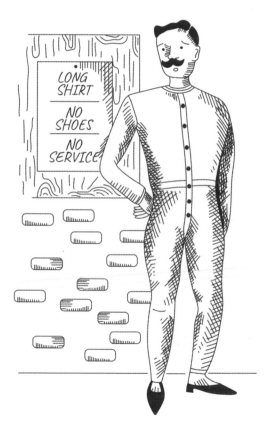

THE T-SHIRT ORIGINATED AS LONG JOHN UNDERGARMENTS FOR MEN. THEY WERE INITIALLY TABOO TO BE WORN ON THEIR OWN, AND THERE WERE EVEN LAWS IN 1890s HAVANA BANNING THEM FROM BEING WORN IN PUBLIC.

LONG SHIRT

NO SHOES

NO SERVICE

THIS ATTITUDE SOON CHANGED. **COOPER UNDERWEAR COMPANY** RELEASED AN AD IN 1904 MARKETING T-SHIRTS TOWARD **BACHELORS WHO DIDN'T HAVE WIVES TO SEW AND MEND THEIR BUTTONS.**

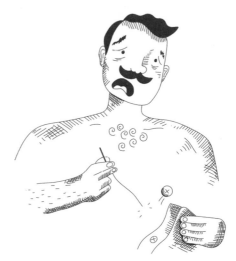

SOON AFTER, THE **UNITED STATES NAVY** INTEGRATED THE T-SHIRT INTO THEIR UNIFORM, AND IT GREW TO BECOME ONE OF THE STAPLES OF AMERICAN DRESS.

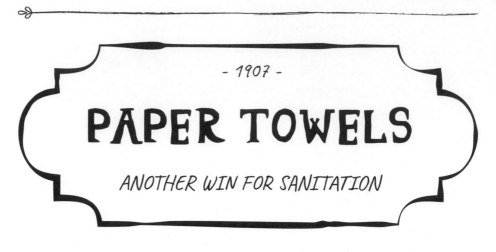

PAPER TOWELS

- 1907 -

ANOTHER WIN FOR SANITATION

SCOTT PAPER COMPANY INVENTED SANI-TOWELS IN 1907 AS **A MORE SANITARY ALTERNATIVE** TO CLOTH TOWELS IN PUBLIC RESTROOMS.

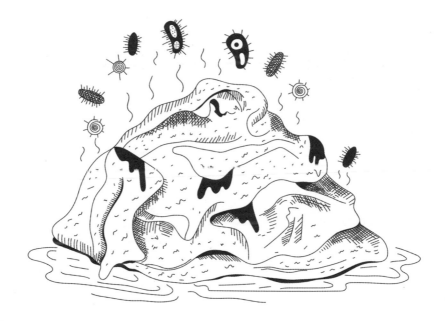

THE INVENTION IS SAID TO HAVE BEEN INSPIRED BY A **PHILADELPHIA SCHOOL TEACHER** WHO HANDED OUT SOFT PAPER TO HER STUDENTS SO THEY WOULDN'T SPREAD THE COMMON COLD ON THE SHARED TOWELS AT SCHOOL.

THE FIRST DISPOSABLE PAPER TOWELS, SANI-TOWELS WERE FAR MORE SANITARY FOR PUBLIC USE AND DIDN'T REQUIRE WASHING SERVICES FOR COMMUNAL BATHROOM TOWELS. THEY'VE SURPASSED THEIR ORIGINAL USE AS COLD-FIGHTERS TO BE **THE NUMBER ONE CHOICE FOR HOUSEHOLD CLEAN UP,** AND HAVE EVEN REPLACED NAPKINS IN MANY HOMES.

COFFEEMAKERS

COFFEE GROUNDS ARE SO 1800s

MELITTA BENTZ WAS A HOUSEWIFE
IN DRESDEN, GERMANY WHO GREW
ANNOYED BY THE GROUNDS IN HER
COFFEE CUP. GROUNDS WOULD ALSO
SIT SATURATING IN THE COFFEE
POT, MAKING IT MORE BITTER
AS TIME WENT ON, AND WERE
A PAIN TO CLEAN OUT AFTER
THE FACT. SEPARATING THEM
WAS A NATURAL MOVE, AND
BENTZ DID JUST THAT WHEN
SHE INVENTED THE COFFEE
FILTER IN 1908.

WHILE EXPERIMENTING WITH DIFFERENT METHODS, **SHE GRABBED A PIECE OF BLOTTING PAPER FROM HER SON'S SCHOOL NOTEBOOK AND USED IT TO FILTER THE COFFEE.** HER INVENTION MADE COFFEE MORE ENJOYABLE, AND MINIMIZED MESSY CLEAN UP WITH A QUICKER, DISPOSABLE OPTION.

Hot Water →

Coffee

← Blotting Paper

← Brass Pot With Holes

Cup →

12
10
8

Coffee first then talk to me ♥

THANKS TO BENTZ, THE **DRIP COFFEEMAKER** WAS MADE POSSIBLE. WITH THE EASE OF DISPOSABLE PAPER FILTERS OVER LINENS THAT REQUIRED TEDIOUS WASHING, THE DRIP COFFEEMAKER BECAME THE EASIEST AND MOST COMMON METHOD FOR BREWING A SIMPLE CUP OF COFFEE.

- 1911-
EGG CARTONS
A PACKAGING TOUR DE FORCE

LEGEND HAS IT THAT IN 1911, A HOTEL OWNER HAD A DISPUTE WITH A LOCAL FARMER WHOSE **EGGS OFTEN ARRIVED BROKEN UPON DELIVERY.** UP UNTIL THAT POINT, EGGS WERE TYPICALLY CARRIED IN EGG BASKETS.

JOURNALIST AND NEWSPAPER PUBLISHER **JOSEPH COYLE** DECIDED TO TAKE ON THE PROBLEM, AND INVENTED THE FIRST EGG CARTON, **THE COYLE "EGG-SAFETY CARTON."** HIS DESIGN FEATURED A CARTON MADE OF CARBOARD OR PAPER WITH INDIVIDUAL COMPARTMENTS THAT HAD CUSHIONING FOR THE EGGS. HE MADE HIS FIRST CARTONS BY HAND, AND WENT ON TO INVENT A MACHINE TO PRODUCE THEM.

EGGS RAYLITE EGG BOX

HANDLE WITH CARE

EGGS-TREME COMFORT. SO COZY!

BUT LIKE MOST INVENTIONS, THE EGG CARTON WASN'T A ONE-HIT WONDER. BEFORE COYLE'S INVENTION, A MAN IN LIVERPOOL NAMED **THOMAS PETER BETHELL** INVENTED THE "RAYLITE EGG BOX" IN 1906. THE BOXES FEATURED INTERLOCKING CARDBOARD OR WOOD INSIDE TO HOLD THE EGGS, BUT WERE TYPICALLY SIZEABLE WOODEN BOXES WITH A HANDLE, UNLIKE COYLE'S MORE COMPACT SOLUTION.

- 1919 -

TOASTERS

*ACTUALLY THE BEST THING
BEFORE SLICED BREAD*

THE POP-UP TOASTER WAS
INVENTED BY A MECHANIC NAMED
CHARLES STRITE WHO NOTICED
HOW OFTEN TOAST GOT BURNT IN
WORLD WAR I CAFETERIAS.

TOASTER
2000

EARLIER MODELS OF THE TOASTER FEATURED **EXPOSED WIRES** AND HAD **NO TIMER** TO TELL WHEN THE BREAD WAS DONE. A PERSON HAD TO WATCH AND MANUALLY FLIP THE SLICE, AS THE MACHINES ONLY TOASTED ONE SIDE AT A TIME.

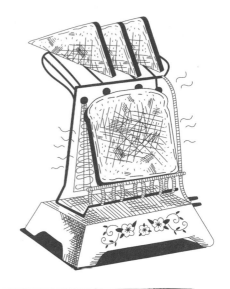

STRITE'S DESIGN TOASTED **BOTH SIDES OF THE BREAD AT ONCE AND USED A SPRING AND TIMER TO POP UP THE BREAD WHEN IT WAS READY.** HIS TOASTER TOOK OFF EVEN MORE ONCE SLICED BREAD WAS INVENTED IN 1928, AND IT REMAINS ONE OF THE MOST COMMON HOUSEHOLD APPLIANCES TODAY.

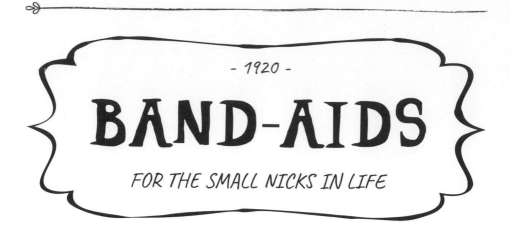

BAND-AIDS

FOR THE SMALL NICKS IN LIFE

IT WAS A NEW BRUNSWICK, NEW JERSEY, COUPLE NAMED **EARLE** AND **JOSEPHINE DICKSON** WHO SAVED US ALL FROM OUR PAPERCUTS BY INVENTING THE BAND-AID. JOSEPHINE WOULD OFTEN GET SMALL CUTS AND BURNS WHILE COOKING AND WRAPPING THEM ALONE WAS CHALLENGING AND INCONVENIENT.

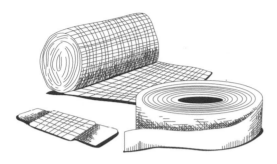

EARLE WORKED FOR **JOHNSON & JOHNSON,** AND HE AND JOSEPHINE USED THE BRAND'S GAUZE AND ADHESIVE TAPE TO CREATE PRE-ROLLED STRIPS OF TAPE WITH THIN PADS OF GAUZE.

THE DICKSONS BROUGHT THEIR INVENTION TO THE COMPANY, AND THE DESIGN HASN'T CHANGED MUCH SINCE. **BAND-AIDS** CORNERED THE MARKET ON QUICK FIXES FOR SMALL, EVERYDAY WOUNDS, AND ARE STILL COMMON IN EVERY HOUSEHOLD FIRST-AID KIT TODAY.

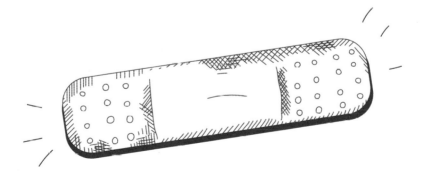

PASSPORTS

PROVING YOUR FACE IS YOUR FACE

Brown Hair ✓
Brown Eyes ✓
Liquid in 3 oz ✓

THE PASSPORT WASN'T ALWAYS A STRICT NECESSITY FOR TRAVEL. EARLY VERSIONS IN THE LATE 1800s TO EARLY 1900s FEATURED **WRITTEN OUT, PHYSICAL DESCRIPTIONS OF A PERSON'S FACE.**

WHEN PHOTOGRAPHS WERE FIRST ADDED, THERE WERE NO REQUIREMENTS. MANY PEOPLE SUBMITTED PHOTOS OF THEMSELVES THEY FOUND ATTRACTIVE, SOMETIMES EVEN ENJOYING **A FAVORITE HOBBY.**

IT WAS IN 1920 THAT **THE LEAGUE OF NATIONS STANDARDIZED PASSPORTS WORLDWIDE,** THOUGH THEY WEREN'T WITHOUT THEIR ISSUES. UNTIL 1937, MARRIED WOMEN WERE LISTED ON THEIR HUSBANDS' PASSPORTS, AND THEY WERE NOT ALLOWED TO TRAVEL WITHOUT THEIR HUSBANDS. THAT AND A RISE IN IMMIGRATION CONTROL FORESHADOWED FUTURE STRUGGLES.

TRAFFIC LIGHTS

STOP ON YELLOW OR STEP ON IT

THE **T-SHAPED TRAFFIC SIGNAL** WAS INVENTED BY **GARRETT MORGAN** IN 1923. IT WASN'T THE FIRST TRAFFIC LIGHT, BUT IT HAD ONE THING THE OTHERS DIDN'T: A THIRD SIGNAL OTHER THAN "STOP" AND "GO."

MORGAN WAS THE SON OF TWO FORMER SLAVES. HE ALSO INVENTED THE SMOKE HOOD (AN EARLY GAS MASK) AND STARTED ONE OF THE MOST SIGNIFICANT BLACK NEWSPAPERS IN THE UNITED STATES, THE CLEVELAND CALL AND POST.

STOP

SLOW DOWN PLEASE! HIT THE BRAKES!

OK, OK GO AHEAD!

THE ELECTRIC TRAFFIC SIGNAL HAD BEEN INVENTED IN 1914, BUT THERE WERE STILL MANY ACCIDENTS BECAUSE DRIVERS HAD NO TIME TO REACT TO THE SUDDEN SWITCH FROM STOP TO GO. MORGAN DESIGNED A SIGNAL WITH **A WARNING POSITION** IN BETWEEN, AND SOLD HIS INVENTION TO GENERAL ELECTRIC FOR $40,000.

- 1928 -

SLICED BREAD

THE BEST THING

THE BREAD-SLICING MACHINE WAS INVENTED IN 1928 BY **OTTO ROHWEDDER**, AND AT FIRST IT WAS CONSIDERED FAR FROM GREAT.

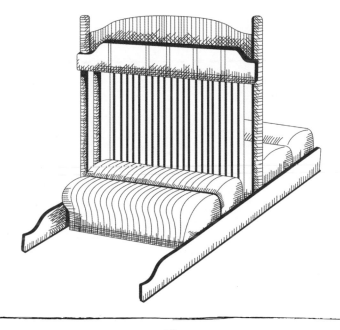

CONSUMERS WERE WARY OF THE PRE-SLICED BREAD BECAUSE OF ITS UNAPPEALING LOOK AND TENDENCY TO GO STALE FASTER THAN A WHOLE LOAF.

SLICED BREAD
Eat Immediately

FRESH BREAD
Baked FRESH DAILY!

AFTER ADDING **U-SHAPED PINS** TO HOLD THE LOAF TOGETHER TO MAINTAIN FRESHNESS, THE CONVENIENCE OF SLICED BREAD WON OUT, AND BAKED GREATNESS WAS ACHIEVED.

PONDER BREAD
SLICED!

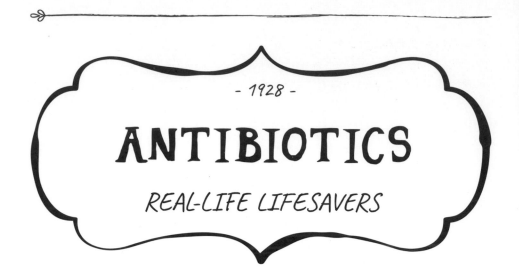

- 1928 -

ANTIBIOTICS

REAL-LIFE LIFESAVERS

THE ERA OF ANTIBIOTICS BEGAN WITH THE SOMEWHAT ACCIDENTAL DISCOVERY OF PENICILLIN BY **MICROBIOLOGIST ALEXANDER FLEMING.** HE FOUND THAT A PETRI DISH CONTAINING A SAMPLE OF THE BACTERIA STAPHYLOCOCCUS HAD BEEN CONTAMINATED BY THE FUNGUS, **PENICILLIUM NOTATUM.** UPON EXAMINATION, HE REALIZED THAT THE BACTERIA HAD BEEN ELIMINATED WHERE THE MOLD GREW.

BEFORE ANTIBIOTICS, IT WASN'T SO MUCH THE INJURIES, BUT THE BACTERIAL INFECTIONS THAT FOLLOWED THAT CLAIMED SO MANY LIVES, PARTICULARLY IN WAR. WITH NO WAY TO COMBAT THESE INFECTIONS, THE MAJORITY OF BATTLEFIELD WOUNDS PROVED **FATAL**.

FLEMING ISOLATED AND STUDIED THE PENICILLIUM MOLD, DISCOVERING ITS ABILITY TO FIGHT BACTERIA, AND ITS POTENTIAL TO DO THE SAME WITH DISEASE. BY THE 1940s, PENICILLIN WAS MASS-PRODUCED AS AN ANTIBIOTIC. **IT SAVED COUNTLESS LIVES IN WORLD WAR II,** DROPPING THE RATE OF DEATH FROM BACTERIAL PNEUMONIA FROM 18 PERCENT IN WORLD WAR I TO 1 PERCENT IN WORLD WAR II.

- 1928 -

LEATHER JACKETS

"I'M COOLER THAN YOU"

THE LEATHER JACKET ORIGINATED AS **BOMBER JACKETS WORN BY FIGHTER PILOTS DURING WORLD WAR I.** THE LEATHER WAS HEAVILY LINED AND KEPT THEM WARM AT HIGH ALTITUDES.

IN 1928, **IRVING SCHOTT** DESIGNED THE FIRST LEATHER JACKET WITH A ZIPPER INSTEAD OF A BUTTON-FRONT, MAKING IT THE IDEAL MOTORCYCLE JACKET. HE NAMED IT AFTER A CIGAR, THE PERFECTO, AND SOLD IT AT HARLEY DAVIDSON FOR ONLY $5.50. IT ACHIEVED CULT STATUS AFTER **MARLON BRANDO** WORE IT IN THE WILD ONE.

SOON THE LEATHER JACKET BECAME AN ICONIC PIECE, STYLED INFINITE WAYS, FROM THE UNIFORM OF THE **BLACK PANTHERS** TO THE SIGNATURE PUNK LOOK OF **THE RAMONES.**

SCOTCH TAPE

THE QUICK FIX

OH 3M! Why don't you hire me! I've come from Minnesota with a banjo on my knee!

A YOUNG BANJO PLAYER AND ENGINEERING SCHOOL DROPOUT NAMED **RICHARD GURLEY DREW** WROTE AN ENDEARING LETTER TO THE MINNESOTA MINING AND MANUFACTURING CO. (3M), ASKING FOR COMMERCIAL WORK AND EXPERIENCE IN EXCHANGE FOR ANY SALARY THEY SAW FIT. THE COMPANY MAINLY PRODUCED SANDPAPER AT THE TIME. BUT IT WASN'T LONG UNTIL THIS UNLIKELY CANDIDATE ENDED UP STRIKING INVENTION GOLD.

WHILE VISITING AUTOBODY SHOPS, DREW WITNESSED THE FRUSTRATION OF AUTOMAKERS TRYING TO PAINT **TWO-TONED CARS,** WHICH WERE HUGELY POPULAR IN THE 1920s. IN ORDER TO COVER ONE PORTION OF THE CAR WHILE PAINTING THE OTHER, WORKERS HAD TO USE NEWSPAPER, HOMEMADE GLUE, OR HEAVY SURGICAL TAPES, WHICH CHIPPED THE PAINT JOB WHEN REMOVED.

WITHIN A FEW YEARS, DREW HAD DEVELOPED A FORMULA FOR **GENTLE BUT EFFECTIVE ADHESIVE TAPE.** WITH THE INVENTION OF CLEAR CELLOPHANE, THEY DECIDED TO MAKE THE TAPE TRANSPARENT, AND 3M'S SCOTCH BRAND CELLULOSE TAPE WAS RELEASED IN 1930.

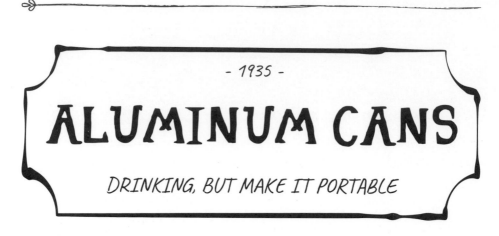

ALUMINUM CANS

DRINKING, BUT MAKE IT PORTABLE

BEFORE CANS, DRINKS WERE MOSTLY SERVED IN **GLASS BOTTLES**, BUT BOTTLES WERE MORE CHALLENGING TO STACK, TOOK LONGER TO CHILL, AND WERE MORE ACCIDENT-PRONE. IN 1935, THE GOTTFRIED KRUEGER BREWING COMPANY AND THE AMERICAN CAN COMPANY RAN THE FIRST TEST OF 2,000 CANS IN RICHMOND, VIRGINIA. THEY WERE MET WITH AN INCREDIBLY POSITIVE RESPONSE, AND CAN PRODUCTION TOOK OFF.

THE FIRST CANS HAD EITHER **FLAT TOPS** YOU WOULD MAKE HOLES IN WITH A SO-CALLED CHURCH KEY, OR **CONE-TOPS,** WHICH WERE SHAPED LIKE THE TOPS OF BOTTLES SO THAT SMALLER BREWERIES COULD CONTINUE USING THE SAME EQUIPMENT.

IN 1962, THE **PULL-TAB CAN** WAS INVENTED, BUT IT CAUSED A HUGE AMOUNT OF LITTER WITH THE DISCARDED TABS. IT WAS MODIFIED TO THE **STAY-TAB TOP** WE KNOW TODAY.

BALLPOINT PENS

LET'S GET THIS INK ROLLING

A HUNGARIAN JOURNALIST NAMED **LÁSZLÓ BÍRÓ** NOTICED THAT NEWSPAPER INK DRIED MUCH FASTER THAN THE SMUDGE-PRONE LIQUID INK OF FOUNTAIN PENS.

This is very Drippy

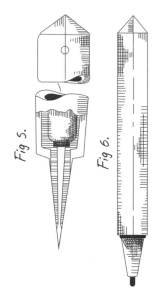

Fig 5.

Fig 6.

THE NEWSPAPER INK WAS TOO THICK TO DRIP THROUGH THE NIB OF THE PEN, INSPIRING BÍRÓ TO USE A **TINY METAL BALL** AT THE OPENING OF THE PEN, WHICH WOULD ROLL THE INK ONTO THE PAGE.

BALLPOINT PENS EVENTUALLY TOOK OVER IN POPULARITY. **THEY HAD A MUCH LONGER INK-LIFE BEFORE NEEDING A REFILL, WHICH MADE WRITING NEATLY A LOT EASIER.**

- 1938 -

CHOCOLATE CHIP COOKIES

AMERICA'S FAVORITE SWEET

WE HAVE A WOMAN FROM MASSACHUSETTS NAMED **RUTH WAKEFIELD** TO THANK FOR MILK'S ULTIMATE COMPANION. SHE INVENTED THE CHOCOLATE CHIP COOKIE WHILE RUNNING HER **TOLL HOUSE INN RESTAURANT.**

YOU'RE WELCOME

LEGEND HAS IT THAT WHILE BAKING COOKIES, WAKEFIELD RAN OUT OF SQUARE BAKER'S CHOCOLATE TO MELT, AND QUICKLY CHOPPED UP PIECES OF A **NESTLÉ CHOCOLATE BAR** TO USE INSTEAD.

BUT WAKEFIELD WAS A TRAINED CHEF, AND IT IS MORE LIKELY THAT SHE INVENTED THE RECIPE INTENTIONALLY. IT WAS PUBLISHED IN **RUTH WAKEFIELD'S TRIED AND TRUE COOKBOOK** IN 1938, AND A YEAR LATER NESTLÉ MADE A DEAL (REPORTEDLY FOR ONLY $1 AND FREE CHOCOLATE FOR LIFE) TO PRINT HER RECIPE ON THE BACK OF THEIR SEMISWEET CHOCOLATE PACKAGES.

COOKIE RECIPE

Original Nestlé Toll House
CHOCOLATE CHIP
COOKIES

We really lucked out
with this recipe.

We got it for such a
great price, let me
tell you.

TRY TO SHARE

- 1943 -

DUCT TAPE

NOT JUST FOR DUCKS OR DUCTS

DUCT TAPE WAS INVENTED BY **VESTA STOUDT,** A MOTHER WHOSE TWO SONS WERE IN THE UNITED STATES NAVY. WHILE SHE WAS WORKING AT THE GREEN RIVER ORDNANCE PLANT IN ILLINOIS, STOUDT PACKAGED BOXES OF AMMUNITION CARTRIDGES, WHICH WERE SHUT WITH PAPER TAPE, DIPPED IN WAX FOR WATERPROOFING, AND PULLED OPEN FROM A SMALL TAB. THE TAB WOULD OFTEN FALL OFF, MAKING THE BOXES DIFFICULT TO OPEN FOR SOLDIERS RELOADING DURING BATTLE.

IN 1943, **STOUDT WROTE A LETTER TO PRESIDENT ROOSEVELT** DESCRIBING THE NEED FOR A DURABLE, CLOTH-BASED, WATERPROOF TAPE. PRESIDENT ROOSEVELT FORWARDED THE LETTER TO THE WAR PRODUCTION BOARD, WHO APPROVED HER IDEA AND HAD JOHNSON & JOHNSON PRODUCE THE TAPE.

Yes, Johnson & Johnson? We have an order for you.

IT WAS KNOWN AS **"DUCK TAPE"** BECAUSE IT WAS WATERPROOF, LIKE "WATER OFF A DUCK'S BACK." SOME SAY IT NATURALLY EVOLVED TO "DUCT TAPE" OR WAS RENAMED WHEN IT WAS USED TO REPAIR AIR DUCTS AFTER THE WAR.

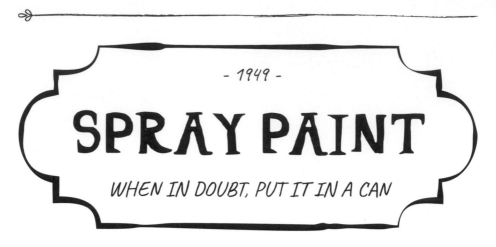

SPRAY PAINT

WHEN IN DOUBT, PUT IT IN A CAN

A PAINT SALESMAN NAMED **EDWARD SEYMOUR** FROM SYCAMORE, ILLINOIS, WAS LOOKING FOR A WAY TO DISPENSE HIS NEW ALUMINUM RADIATOR PAINT. HIS WIFE BONNIE SUGGESTED USING A SPRAY CAN.

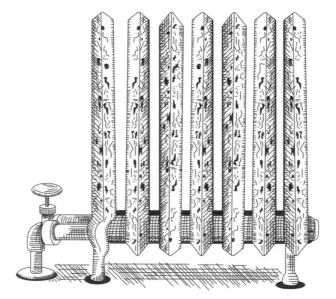

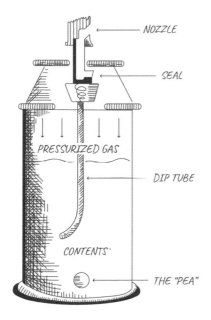

NOZZLE

SEAL

PRESSURIZED GAS

DIP TUBE

CONTENTS

THE "PEA"

SEYMOUR FILLED AN AEROSOL CAN WITH HIS PAINT. **A SMALL METAL BALL INSIDE CALLED THE "PEA" MIXED THE PAINT WHEN SHAKEN, AND THE SPRAY NOZZLE GAVE IT AN EVEN APPLICATION.**

THANKS SEYMOUR

REALIZING THE CONVENIENCE AND EASE OF HIS NEW PAINTING METHOD, SEYMOUR'S COMPANY **"SEYMOUR OF SYCAMORE"** BEGAN MANUFACTURING ALL KINDS OF AEROSOL SPRAY PAINT AND GAVE FUTURE GRAFFITI ARTISTS THEIR WEAPON OF CHOICE.

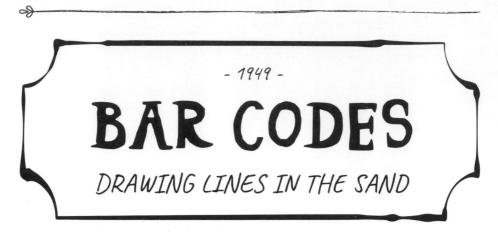

BAR CODES

DRAWING LINES IN THE SAND

IN 1949, A DREXEL INSTITUTE GRADUATE STUDENT NAMED **JOSEPH WOODLAND** INVENTED THE BAR CODE BY DRAWING HIS FINGERS THROUGH THE SAND ON MIAMI BEACH.

HE WANTED TO FIND A WAY TO HURRY CHECK OUT AND STOCKTAKING AT GROCERY STORES, AND INSPIRED BY MORSE CODE, DREW A SERIES OF LINES IN THE SAND. HE REALIZED THAT **BY VARYING THE WIDTHS OF THE LINES,** SIMILAR TO THE SYSTEM OF DOTS AND DASHES, **HE COULD RECORD A UNIQUE CODE FOR EACH ITEM.**

SCAN IT THEN

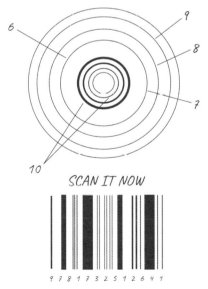

SCAN IT NOW

9 7 8 1 7 3 2 5 1 2 6 4 1

HIS FIRST DESIGN FEATURED LINES AS WELL AS A **BULL'S-EYE-SHAPED BAR CODE,** AND IN 1974 AT A SUPERMARKET IN OHIO, THE UNIVERSAL BAR CODE WAS SCANNED FOR THE FIRST TIME.

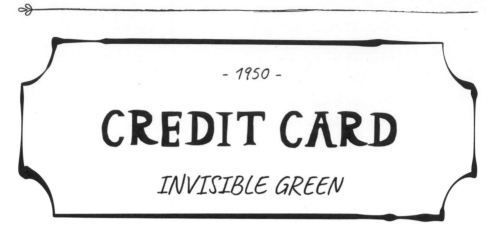

CREDIT CARD

INVISIBLE GREEN

THE FIRST MULTIPURPOSE CHARGE CARD WAS THE DINERS CLUB CARD, FOUNDED BY **FRANK MCNAMARA** AFTER A CLIENT MEAL WHERE HE REALIZED HE'D LEFT HIS WALLET IN ANOTHER SUIT.

IT WASN'T THE FIRST CREDIT CARD EVER MADE, BUT IT WAS THE FIRST TO BE ACCEPTED BY MULTIPLE MERCHANTS. CHARGE PLATES AND CARDS WERE ISSUED AT SOME DEPARTMENT STORES AND OIL COMPANIES BUT COULD ONLY BE USED AT THEIR RESPECTIVE BUSINESSES.

SIGN UP TODAY!

PROVIDENCE
CHARGE-A-PLATE
ASSOCIATES

COMES WITH HANDY CARRYING CASE!

Use Your Charge-A-Plate!
Sign here in ink

THE DINERS CLUB CARD REQUIRED ONE PAYMENT ON A MONTHLY BASIS AND MADE A PROFIT BY CHARGING A 7 PERCENT FEE FROM PARTICIPATING BUSINESSES. MCNAMARA MARKETED IT TO CONSUMERS AS A METHOD OF CONVENIENCE AND EXCLUSIVITY, AND TO MERCHANTS BY GUARANTEEING **CUSTOMERS WITH CARDS WOULD SPEND MORE THAN THOSE WITHOUT.**

TREAT yoself

SALES

DINER'S CLUB CARD

CASH

MONTH

- 1966 -

HAND SANITIZER

*BECAUSE WASHING HANDS
TAKES WAY TOO LONG*

GOOP
For Hands

IN 1966, A NURSING STUDENT NAMED **LUPE HERNANDEZ** WAS LOOKING FOR A FASTER WAY TO CLEAN HANDS THAN WITH SOAP AND WATER. **SHE REALIZED THAT A GEL FORM OF ALCOHOL COULD ELIMINATE GERMS AND BACTERIA, AND PATENTED HER INVENTION.**

HOWEVER, HERNANDEZ WASN'T THE ONLY ONE WORKING ON THE HAND SANITIZER GAME. A COUPLE IN AKRON, OHIO, NAMED **JERRY AND GOLDIE LIPPMAN** POPULARIZED AN ALCOHOL-FREE HAND CLEANER IN 1946. GOLDIE WORKED AT A RUBBER FACTORY, AND THE ONLY WAY TO REMOVE CARBON AND TAR FROM HANDS WAS WITH HARSH CHEMICAL CLEANERS. THE LIPPMANS INVENTED A HEAVY-DUTY CLEANSER THAT WAS GENTLER ON THE SKIN. THEY NAMED IT GOJO.

GOJO WAS POPULAR IN CAR GARAGES BUT BECAME EXPENSIVE WHEN MECHANICS KEPT TAKING IT HOME. THAT'S WHEN JERRY LIPPMAN INVENTED THE **WALL-MOUNTED DISPENSER**, WHICH BECAME POPULAR DUE TO ITS CONVENIENCE. AND 36 YEARS LATER, GOJO INVENTED PURELL.

- 1967 -

CALCULATORS

2 + 2 *HAS NEVER BEEN EASIER*

FROM **THE ABACUS OF ANCIENT SUMER** TO THE FIRST MECHANICAL CALCULATOR IN 1642, HUMANS ALWAYS SOUGHT A WAY TO MAKE MATHEMATICS FASTER AND MORE ACCESSIBLE. IT WASN'T UNTIL **TEXAS INSTRUMENTS** INVENTED THE HANDHELD CALCULATOR IN 1967 THAT OUR ACCESS TO CALCULATORS BECAME UNIVERSAL.

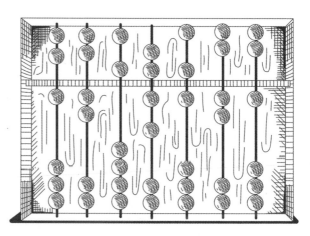

TRY TO CARRY YOURS HOME TODAY!

THE ELECTRONIC CALCULATOR ALREADY EXISTED, BUT BEFORE THE ADVENT OF INTEGRATED CIRCUIT CHIPS, EARLIER MODELS COULD WEIGH AROUND **50 POUNDS** AND WERE PROHIBITIVELY EXPENSIVE.

THE DRAW OF THE **HANDHELD CALCULATOR** WAS THAT IT WAS PORTABLE AND COULD BE USED ANYWHERE. FROM THERE, CALCULATORS BECAME AVAILABLE FOR EVERY STUDENT AND ON EVERY SMARTPHONE.

POST-IT NOTES

A STICKY REMINDER

IN THE LATE-1960s, SCIENTIST **DR. SPENCER SILVER** WAS TASKED WITH DEVELOPING A SUPER-STRONG ADHESIVE. WHAT HE DISCOVERED INSTEAD WAS A REUSABLE ADHESIVE THAT STUCK TO SURFACES GENTLY, NOT PERMANENTLY.

SILVER'S COLLEAGUE **ART FRY** SUGGESTED USING THE STUFF TO STICK BOOKMARKS INTO HYMN BOOKS AT CHURCH. HE HAD ALWAYS WANTED TO PREVENT THEM FROM FALLING OUT WITHOUT RUINING THE PAGES.

WHILE THEY WERE DESIGNING THE PROTOTYPE, THE LAB NEXT DOOR ONLY HAD YELLOW SCRAPS OF PAPER. THE COLOR STUCK, AND **OFFICE LIFE WAS NEVER THE SAME AGAIN!**

- 1971 -

ICE PACKS

HOT AND COLD YOU CAN CARRY

THE GEL ICE-PACK WAS INVENTED BY A LONG ISLAND PHARMACIST NAMED **JACOB SPENCER** IN 1971. SPENCER'S ICE-PACK WAS ALSO THE FIRST ONE DESIGNED TO WORK FOR THE BODY.

PREVIOUS LIQUID OR ALCOHOL-BASED ICE-PACKS HAD **A TENDENCY TO LEAK OR DEVELOP MOLD,** AND THE FIRST INSTANT COLD PACK INVENTED BY **ALBERT A. ROBBINS** IN 1959 WAS ONLY ABLE TO KEEP FOOD AND DRINKS COLD. IT WASN'T MADE FOR THE BODY, OR TO BE USED FOR HEAT AS WELL.

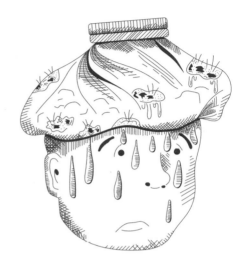

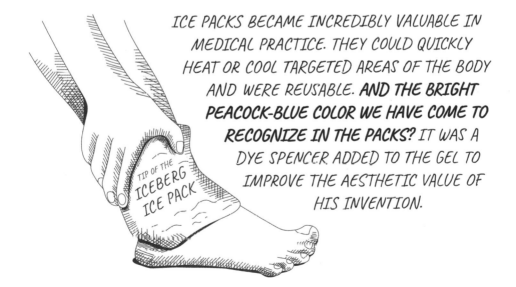

ICE PACKS BECAME INCREDIBLY VALUABLE IN MEDICAL PRACTICE. THEY COULD QUICKLY HEAT OR COOL TARGETED AREAS OF THE BODY AND WERE REUSABLE. **AND THE BRIGHT PEACOCK-BLUE COLOR WE HAVE COME TO RECOGNIZE IN THE PACKS?** IT WAS A DYE SPENCER ADDED TO THE GEL TO IMPROVE THE AESTHETIC VALUE OF HIS INVENTION.

- 1972 -

BREAKFAST SANDWICHES

THE ULTIMATE MEAL TO GO

BREAKFAST SANDWICHES CAN BE TRACED BACK TO 19TH CENTURY LONDON, WHERE BUSY FACTORY WORKERS COULD GRAB THEM FROM STREET STANDS ON THE WAY TO WORK. THEY WERE CALLED **"BAP" SANDWICHES** AFTER THE ROLLS THEY WERE SERVED IN.

WITH THE **INDUSTRIAL REVOLUTION**, A HOT, FULL BREAKFAST THAT COULD BE WRAPPED UP INTO ONE QUICK PACKAGE BECAME A DAILY STAPLE FOR WORKERS. *ITS POPULARITY SOARED AND THE FIRST COMMERCIAL BREAKFAST SANDWICH WAS BORN IN 1972 WHEN MCDONALD'S CREATED THE EGG MCMUFFIN.*

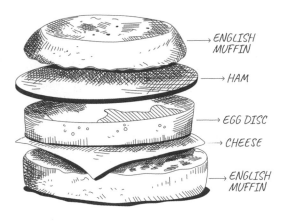

→ ENGLISH MUFFIN

→ HAM

→ EGG DISC

→ CHEESE

→ ENGLISH MUFFIN

HERB PETERSON WAS TRYING TO MAKE A VERSION OF EGGS BENEDICT THAT DIDN'T REQUIRE HOLLANDAISE SAUCE, AND HE ENDED UP WITH THE EGG MCMUFFIN. IT BECAME ONE OF MCDONALD'S MOST POPULAR ITEMS AND MADE FAST FOOD BREAKFAST A HIT.

GPS

IT FINDS EVERYTHING

SATELLITE

REFERENCE
ELLIPSOID

YOU'LL NEVER BE LOST AGAIN

IN THE EARLY 1970s A TEAM OF SCIENTISTS AND ENGINEERS LED BY **ROGER EASTON** INVENTED THE GLOBAL POSITIONING SYSTEM (GPS). HOWEVER, IT WAS MATHEMATICIAN **GLADYS WEST** WHO TRACKED AN ULTRA-ACCURATE SURFACE MODEL OF THE EARTH'S SHAPE. GLADYS AND HER TEAM PROGRAMMED IT INTO SUPERCOMPUTERS, CREATING THE TECHNOLOGY THAT WAS INSTRUMENTAL TO GPS.

WEST'S CALCULATIONS, DATA PROCESSING, AND PROGRAMMING ENSURED THE MOST CRUCIAL PART OF GPS: ITS **ACCURACY.** IT IS NOW USED FOR CELL PHONES, CARS, WEATHER FORECASTING, SOCIAL MEDIA, FARMING, TRACKING ENDANGERED SPECIES, AND A MYRIAD OF MODERN INDUSTRIES AND TECHNOLOGIES WE RELY ON TODAY.

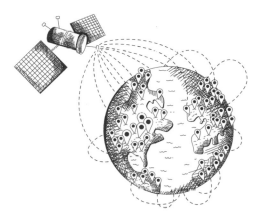

SHE HAD NO IDEA HOW HER WORK WOULD INFLUENCE THE WORLD: **"WHEN YOU'RE WORKING EVERY DAY, YOU'RE NOT THINKING, 'WHAT IMPACT IS THIS GOING TO HAVE ON THE WORLD?' YOU'RE THINKING, 'I'VE GOT TO GET THIS RIGHT.'"** AND SHE DID.

- 1982 -

SUPER SOAKERS

WEAPONIZED FUN

BEFORE THE SUPER SOAKER, **SQUIRT GUNS** SHOT WEAK
SPURTS OF WATER THAT ONLY WENT A SHORT DISTANCE
AND HAD TO BE REFILLED EVERY COUPLE OF BLASTS.

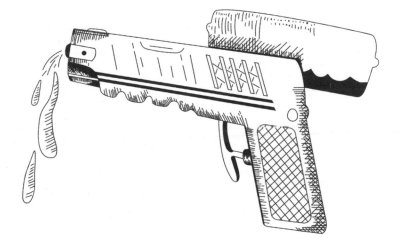

THEN A NASA ENGINEER GOT INVOLVED. **LONNIE JOHNSON** WAS WORKING ON AN ENVIRONMENTALLY FRIENDLY HEAT PUMP THAT USED WATER INSTEAD OF FREON. HE FIXED A NOZZLE TO A BATHROOM FAUCET, AND THE PRESSURIZED WATER SHOT ACROSS THE ROOM.

INSTANTLY, HE REALIZED IT WOULD MAKE A GREAT TOY. HIS INVENTION GREW INTO THE ARSENAL OF SUPER SOAKERS WE HAVE COME TO **KNOW, LOVE, AND RUN SCREAMING FROM.**

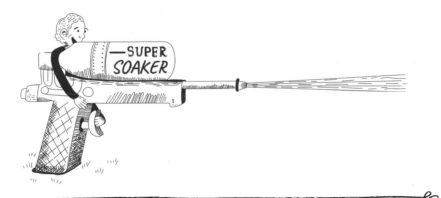

SELFIE STICKS

- 1984 -

DECREASING AWKWARD INTERACTIONS WITH STRANGERS SINCE 1984

A JAPANESE COUPLE NAMED **HIROSHI UEDA** AND **YUJIRO MIMA** WERE VACATIONING IN EUROPE, AND HAVING TROUBLE TAKING PICTURES OF THEMSELVES. INSPIRED BY THEIR PLIGHT, THEY INVENTED THE FIRST SELFIE STICK IN 1984.

THE LOUVRE, 1984

Fig. 17a

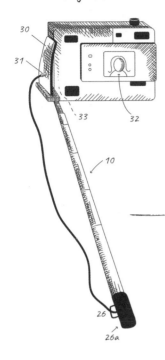

THEY CALLED IT THE
EXTENDER STICK AND THE
ORIGINAL DESIGN INCLUDED
A MIRROR ON THE FRONT
TO HELP THE USER SEE
THEMSELVES.

A PHOTO FROM 1925 WAS RECENTLY
DISCOVERED OF A COUPLE HOLDING
A POLE THAT APPEARS TO BE
ATTACHED TO THE CAMERA. IT'S
POSSIBLE THE IMAGE SHOWS THE
FIRST PROTOTYPE OF WHAT WOULD
BECOME THE SELFIE STICK 60 YEARS
LATER. BY 2014, OVER 100,000 SELFIE
STICKS WERE SOLD IN THE UNITED
STATES IN DECEMBER ALONE.

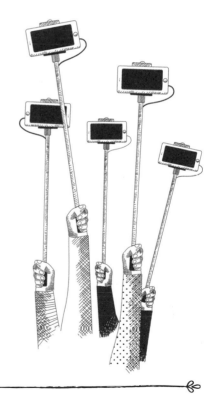

PIZZA SAVERS

NO PIE LEFT BEHIND

IF YOU'RE A CHILD OF THE 1990s, **THE LITTLE WHITE, PLASTIC STANDS** THAT KEEP THE LIDS ALOFT IN PIZZA BOXES ARE BARBIE TABLES, HAMSTER FURNITURE, OR PROJECTILES TO FLICK AT YOUR SIBLINGS.

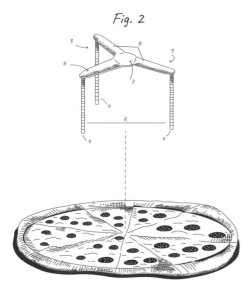

Fig. 2

BUT THE REAL NAME OF THIS QUINTESSENTIALLY '80s INGENIOUS INVENTION IS THE **PIZZA SAVER,** AND IT DOES JUST THAT. IN 1985, **CARMELA VITALE,** A LONG ISLAND MOTHER, GOT A PATENT APPROVED FOR HER "PACKAGE SAVER."

VITALE NOTICED THAT THE HEAT FROM THE PIZZA MADE THE LID OF THE CARDBOARD BOX SAG, STICKING TO THE PIZZA AND RUINING THE CHEESE. SHE DREW UP A PLASTIC PIZZA SAVER, AND THE **SIMPLE, COST-EFFECTIVE SOLUTION STUCK.**

PERFECT CHEESE EVERY TIME

ACKNOWLEDGMENTS

A big thank you to **Allie, Cody,** and **Ryan,** and to my parents, **Jim** and **Kyung,** for inventing me. Thank you to **John Whalen** and **Whalen Book Works** for the opportunity to write this book. My gratitude goes to brilliant illustrator **Rebecca Pry,** designer **Melissa Gerber,** and to **Margaret McGuire Novak,** the best editor and emailer around. And thank you to anyone who decides to carry this little gem on their shelves. I extend my deepest gratitude to all who worked on this book, and my deepest hatred to whomever invented the deadline.

ABOUT THE AUTHOR

Laura Hetherington is a writer, actress, and comedian in New York City from a nice town in Maryland that you've never heard of. She received a BA in English from Fordham University and has a love for *The X-Files,* a good hike, and staring deeply into the eyes of her roommate, Charles the Yorkie.

ABOUT THE ILLUSTRATOR

Rebecca Pry is an Illustrator and Designer living in Warwick, New York. She received a BFA in Illustration from Rhode Island School of Design in 2013. Rebecca's art adds a humorous twist to everyday items and scenes, and she has created patterns and graphics for home goods, books, accessories, apparel, and regularly shows her work in local galleries in the Hudson Valley. When she is not drawing, she is outside in a brightly colored sweater. See more at rebeccapry.com.

ABOUT WHALEN BOOK WORKS

PUBLISHING PRACTICAL & CREATIVE NONFICTION

Whalen Book Works is a small, independent book publishing company based in Kennebunkport, Maine, that combines top-notch design, unique formats, and fresh content to create truly innovative gift books.

Our unconventional approach to bookmaking is a close-knit, creative, and collaborative process among authors, artists, designers, editors, and booksellers. We publish a small, carefully curated list each season, and we take the time to make each book exactly what it needs to be.

We believe in giving back. That's why we plant one tree for every ten books we print. Your purchase supports a tree in the Rocky Mountain National Park.

Get in touch!

Visit us at **WhalenBookWorks.com**
or write to us at
68 North Street, Kennebunkport, ME 04046.

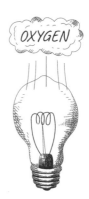

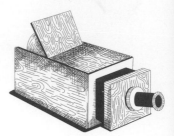